YOUR KNOWLEDGE HAS VALUE

- We will publish your bachelor's and
  master's thesis, essays and papers

- Your own eBook and book -
  sold worldwide in all relevant shops

- Earn money with each sale

Upload your text at www.GRIN.com
and publish for free

Bibliographic information published by the German National Library:

The German National Library lists this publication in the National Bibliography;
detailed bibliographic data are available on the Internet at http://dnb.dnb.de .

Imprint:

Copyright © 2011 GRIN Verlag, Open Publishing GmbH
Print and binding: Books on Demand GmbH, Norderstedt Germany
ISBN: 9783668376694

This book at GRIN:

http://www.grin.com/en/e-book/350486/wind-turbine-aerodynamics

Sandip Kale

Wind Turbine Aerodynamics

GRIN Publishing

AERODYNAMICS OF WIND TURBINE

A wind turbine is a device that extracts kinetic energy of the wind and converts into useful energy. The power produced by a wind turbine depends on the interaction between the wind turbine rotor and the wind. The wind turbine aerodynamics is important, to design a blade and to analyze aerodynamic performance of the rotor. A number of scientists have derived various methods for aerodynamic analysis of wind turbine rotors [BSJB 2001] [MMR 2002] [Pat 2006] [Wood 2011].

1. One dimensional momentum theory

The aerodynamic analysis carried out here is based on momentum theory. For this analysis following assumptions are made,

- Homogenous flow,
- incompressible flow,
- steady state flow,
- No frictional drag,
- A non-rotating wake,
- Infinite number of blades,
- Uniform thrust over rotor area,
- Ambient static pressure at far upstream and far downstream

The control volume to analyze wind turbine aerodynamics is shown in Fig. 1. Surface of stream tube and two cross-sections 1 and 3 of stream tube are the volume boundaries of control volume. The wind flow is across the ends of the stream tube only. To analyze energy extraction process a uniform 'actuator disc' representing wind turbine is considered. This disc creates a pressure discontinuity in the air stream tube flowing through it.

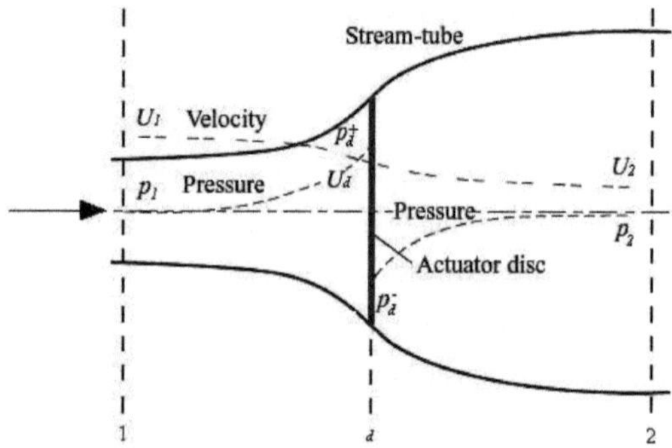

Fig. 1: Actuator disc model of a wind turbine without wake [MMR 2002]

The upstream cross-section area is smaller than that of the disc, while the downstream cross-section area is larger than the disc. The stream-tube is expanding to keep the mass flow rate same everywhere. The mass flow rate along the stream-tube is given as,

$$\rho A_1 U_1 = \rho A_d U_d = \rho A_2 U_2 \tag{1}$$

where ρ is the air density, A is the cross sectional area, U is the air velocity and subscripts indicate values at named cross-sections in Fig. 1.

The actuator disc induces a velocity variation that of the free stream velocity. The stream wise component of this induced flow at the disc is given by aU_1, where a is called the axial flow induction factor, or the inflow factor. The stream wise velocity at the disc is,

$$U_d = (1-a)U_1 \tag{2}$$

When the air passes through the actuator disc, it undergoes an overall change in velocity, $(U_1 - U_2)$. The rate of change of momentum is the product of the overall change of velocity and mass flow rate through the stream-tube,

$$\text{Rate of change of momentum} = (U_1 - U_2)\rho A_d U_d \tag{3}$$

The force causing change of momentum comes completely from the pressure difference across the actuator disc, because the stream-tube is otherwise surrounded by air at atmospheric pressure, which gives zero net force. Therefore,

$$\left(p_d^+ - p_d^-\right)A_d = \left(U_1 - U_2\right)\rho A_d \left(1 - a\right)U_1 \tag{4}$$

The pressure difference in the above equation can be determined by applying Bernoulli's equation. The total energy in upstream and downstream is different. Hence, Bernoulli's equation should be applied separately to the upstream and downstream sections. Therefore, for upstream section,

$$\frac{1}{2}\rho_1 U_1^2 + p_1 + \rho_1 g h_1 = \frac{1}{2}\rho_d U_d^2 + p_d^+ + \rho_d g h_d \tag{5}$$

For incompressible horizontal flow, above equation becomes,

$$\frac{1}{2}\rho U_1^2 + p_1 = \frac{1}{2}\rho U_d^2 + p_d^+ \tag{6}$$

Similarly, for downstream section,

$$\frac{1}{2}\rho U_2^2 + p_2 = \frac{1}{2}\rho U_d^2 + p_d^- \tag{7}$$

Subtracting these equations, we get,

$$\left(p_d^+ - p_d^-\right) = \frac{1}{2}\rho\left(U_1^2 - U_2^2\right) \tag{8}$$

Substituting pressure difference from Equation 8 in Equation 4,

$$\frac{1}{2}\rho\left(U_1^2 - U_2^2\right)A_d = \left(U_1 - U_2\right)\rho A_d \left(1 - a\right)U_1$$

Simplifying and rearranging,

$$U_2 = \left(1 - 2a\right)U_1 \tag{9}$$

The quantity, $U_1 a$, the induced velocity at the rotor, is a combination of the free stream velocity and the induced velocity. As the axial induction factor increase from zero, the wind velocity behind the rotor slows more and more. If $a = 1/2$, the wind has slowed to zero velocity behind the rotor and the simple theory is no longer applicable.

1.1 Power coefficient

The thrust, T can determined by, the net sum of the forces on each side of the actuator disc,

$$T = \left(p_d^+ - p_d^-\right)A_d = \frac{1}{2}\rho A_d U_1^2 4a(1-a) \tag{10}$$

As this thrust is concentrated at the disc, the rate of work done by the thrust is TU_1. Hence, the power extraction from the air is given by,

$$P = TU_1 = \frac{1}{2}\rho A_d U_1^3 4a(1-a)^2 \tag{11}$$

Replacing the control volume area at disc A_d with rotor area A and the free stream velocity U_1 by U the above equation can be written as,

$$P = \frac{1}{2}\rho A U^3 4a(1-a)^2 \tag{12}$$

The power available in the wind for rotor area A is $\frac{1}{2}\rho A U^3$.

The wind turbine performance is usually described by the ratio of rotor power and power in the wind and known as power coefficient, C_P

$$C_P = \frac{P}{\frac{1}{2}\rho A U^3} \tag{13}$$

The fraction of power in the wind that extracted by the rotor is also represented by the non-dimensional power coefficient. From Equation 12, the power coefficient is,

$$C_P = 4a(1-a)^2 \tag{14}$$

The maximum value of power coefficient is given by taking the derivative of the power coefficient with respect to a and equating to zero. Therefore

$$\frac{d}{da}(C_P) = 4(1-a)(1-3a) = 0$$

$$a = \frac{1}{3}$$

Hence, the maximum power coefficient,

$$C_{P_{max}} = \frac{16}{27} = 0.593 \qquad (15)$$

This maximum achievable power coefficient is named as the Betz limit (1926, Albert Betz, the German aerodynamicist). Till date, the wind turbine has not been designed that exceeded the Betz limit [MMR 2002].

Similar to power, the thrust coefficient, C_T, is the ratio of thrust force to dynamic force and mathematically expressed as,

$$C_T = \frac{T}{\frac{1}{2}\rho A U^2} \qquad (16)$$

A non-dimensional thrust coefficient is also represented as,

$$C_T = 4a(1-a) \qquad (17)$$

The thrust coefficient is maximum (C_T=1.0), when a = 0.5 and the downstream velocity is zero. At maximum power output (a = 1/3), C_T has a value of 8/9. A graph of C_P and C_T for an ideal Betz turbine and non-dimensional downstream wind velocity is plotted in Fig. 2.

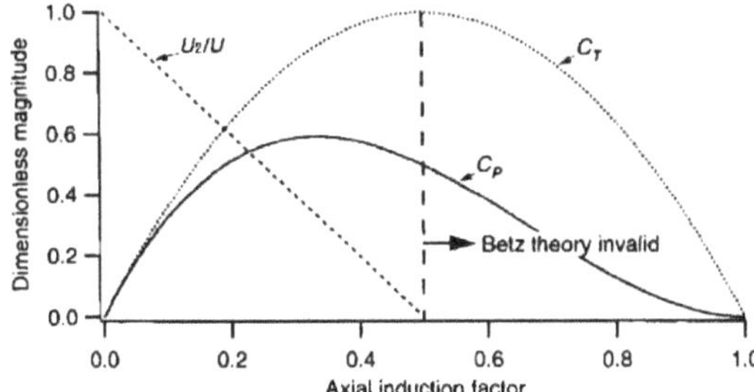

Fig. 2:   Variation of C_P and C_T with axial induction factor a [MMR 2002]

This graph shows, if the axial induction factor is greater than 0.5, this idealized model is not valid. In practice, the thrust coefficient can go as high as two, when the axial induction factor exceeds 1/2. In this case complicated flow patterns obtained and cannot be represented by this simple model.

The Betz limit, 0.593 is the maximum value of the theoretical achievable power coefficient. In practice, this value is further decreased because of wake rotation, finite blade number and related tip losses and aerodynamic drag.

The overall wind turbine efficiency is also dependent on various efficiencies such as mechanical efficiency, electrical efficiency of generator, transmission efficiency, and ineffective area in addition to rotor power coefficient. Hence, overall efficiency of wind turbine is given by,

$$\eta_{overall} = \frac{P_{out}}{\frac{1}{2}\rho A U^3} = \eta_{other} C_P \tag{18}$$

2.   Ideal horizontal axis wind turbine with wake rotation

In practice, wind turbine consists of a rotor with a finite number of blades rotating at an angular velocity Ω about axis parallel to wind direction. In this case the downstream rotates in the opposite direction to the rotor as a result of reaction to the torque exerted by the flow on the wind turbine rotor. Fig. 3 illustrates the annular stream tube model with wake rotation.

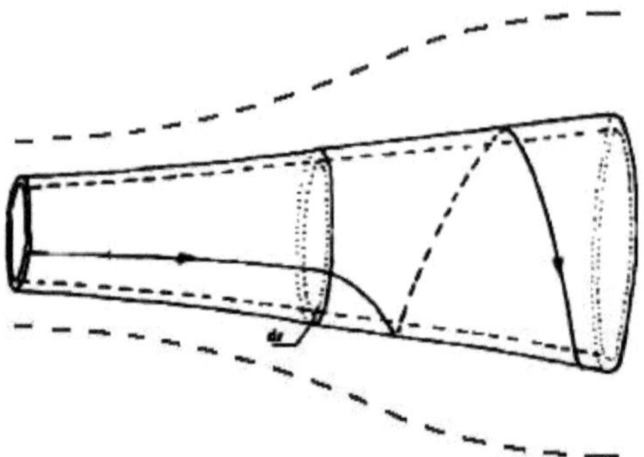

Fig. 3:   Stream tube model with wake rotation [MMR 02]

In this case less energy is extracted by the rotor compared to energy extraction without wake rotation. This is the result of generation of rotational kinetic energy in the wake. Generally, the extra kinetic energy contained in the wake will be higher if the generated torque is higher. Hence, the wake losses are more in wind turbines having slow rotational speeds and higher torque, while less in low torque and high-speed wind turbines.

Fig. 4 illustrates actuator disc model for analysis of wind turbine with wake rotation. The angular velocity imparted to the flow stream, ω is assumed small compare to rotor angular velocity Ω. Consider an annular stream tube of radial thickness dr at radius r. Then its cross-sectional area is equal to $2\pi r dr$. The pressure, wake rotation and induction factors are all assumed to be a function of radius [MMR 02].

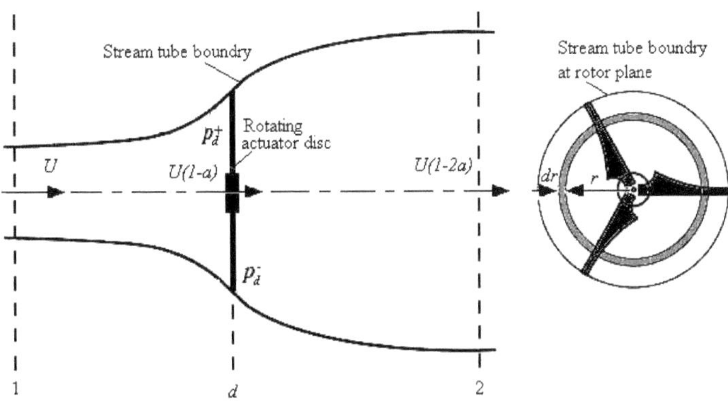

Fig. 4: Actuator disc model of a wind turbine for rotating wake

As control volume moves with an angular velocity of the blades, the pressure difference across the blade can be obtained by applying energy equation before and after the blades. When the air passes the disc, the angular velocity of the air relative to blades increases from (Ω) to $(\Omega + \omega)$, while the axial component of the velocity remains constant.

$$\left(p_d^+ - p_d^-\right) = \rho\left(\Omega - \frac{1}{2}\omega\right)\omega\, r^2 \tag{19}$$

The elemental thrust on an annular element is,

$$dT = \left(p_d^+ - p_d^-\right)dA$$

$$dT = \left[\rho \left(\Omega - \frac{1}{2}\omega \right) \omega r^2 \right] 2\pi dr \tag{20}$$

Defining angular induction factor as,

$$a' = \frac{\omega}{2\Omega} \tag{21}$$

In the analysis with wake rotation the induced velocity at the rotor not only consists of the axial component Ua, but also a component in the rotor plane $r\Omega a'$.

Substituting angular induction factor in the expression of thrust,

$$dT = 4a'\left(1+a'\right)\frac{1}{2}\rho\,\Omega^2 r^2 2\pi dr \tag{22}$$

Similar to Equation 10, the thrust on an annular cross-section can also be expressed as,

$$dT = 4a\left(1-a\right)\frac{1}{2}\rho U^2 2\pi r dr \tag{23}$$

From Equations 2.22 and 2.23,

$$\frac{a\left(1-a\right)}{a'\left(1+a'\right)} = \left(\frac{\Omega\,r}{U}\right)^2 \tag{24}$$

In Equation 24, the term $\Omega\,r$ is the rotor speed at some intermediate radius. The ratio of the rotor speed at intermediate radius to the wind speed is termed as the local speed ratio and denoted as λ_r.

$$\lambda_r = \frac{\Omega\,r}{U} \tag{25}$$

At the blade tip rotor radius is R and Equation 2.25 becomes,

$$\lambda = \frac{\Omega\,R}{U} \tag{26}$$

In above equation λ is the tip speed ratio and obtained as the ratio of the tangential speed at blade tip to the undisturbed wind speed. The relation between the local speed ratio and tip speed ratio is given as,

$$\lambda_r = \lambda \frac{r}{R} \tag{27}$$

The torque exerted on the rotor, Q, is given as change in angular momentum of the wake. For the elemental area,

$$dQ = d\dot{m}(\omega r)r \tag{28}$$

$$dQ = (\rho U_d 2\pi r dr)(\omega r)r \tag{29}$$

Substituting, $U_d = U(1-a)$ and $a' = \omega/2\Omega$ and rearranging,

$$dQ = 4a'(1-a)\frac{1}{2}\rho U \Omega r^2 2\pi dr \tag{30}$$

The power generated at each element, dP, is

$$dP = \Omega\, dQ \tag{31}$$

$$dP = \frac{1}{2}\rho A U^3 \left[\frac{8}{\lambda^2} a'(1-a)\lambda_r^3 d\lambda_r \right] \tag{32}$$

It can be observed that the power from any annular ring is depends on the axial induction factor, angular induction factor and tip speed ratio. Air flow direction and magnitude at the rotor plane can be calculated from induction factors. The local speed ratio is the function of the tip speed ratio and the radius.

The incremental contribution to the power coefficient, dC_P, from each angular ring,

$$dC_P = \frac{dP}{\frac{1}{2}\rho A U^3} \tag{33}$$

The power coefficient is given as,

$$C_P = \frac{8}{\lambda^2} \int_0^\lambda a'(1-a)\lambda_r^3 d\lambda_r \tag{34}$$

Solving Equations 2.24 and 2.25 for a',

$$a' = -\frac{1}{2} + \frac{1}{2}\sqrt{1 + \frac{4}{\lambda_r^2}a(1-a)} \tag{35}$$

From Equation 34, it is clear that the maximum value of the term a' $(1\text{-}a)$ will deliver maximum power. Substituting the value of a' from Equation 35 into the term a' $(1\text{-}a)$ and equating the derivative with respect to a to zero gives,

$$\lambda_r^2 = \frac{(1-a)(4a-1)^2}{(1-3a)} \tag{36}$$

The above equation shows that a is a function of λ_r in each angular ring. From Equation 2.24 and 2.36,

$$a' = \frac{1-3a}{4a-1} \tag{37}$$

The derivative of Equation 36 with respect to a, will give conditions for maximum power production,

$$2\lambda_r d\lambda_r = \left[\frac{6(4a-1)(1-2a)^2}{(1-3a)^2}\right]da \tag{38}$$

Substituting λ_r, a' and $d\lambda_r$ in Equation 34 for maximum value of power coefficient,

$$C_{P_{max}} = \frac{24}{\lambda^2}\int_{a_1}^{a_2}\left[\frac{(1-a)(1-2a)(1-4a)}{1-3a}\right]^2 \tag{39}$$

In this equation limits of integration a_1 and a_2 correspond to axial induction factors, $\lambda_r = 0$ and $\lambda_r = \lambda$, respectively. It gives a lower limit $a_1 = 0.25$ for $\lambda_r = 0$.
From Equation 36,

$$\lambda^2 = \frac{(1-a_2)(1-4a_2)^2}{(1-3a_2)} \tag{40}$$

The value of a_2 can be obtained by solving the above equation for the required tip speed ratio.

By substituting $x = (1\text{-}3a)$ in Equation 39 and evaluating,

$$C_{P_{max}} = \frac{8}{729\lambda^2}\left\{\frac{64}{5}x^5 + 72x^4 + 124x^3 + 38x^2 - 63x - 12(\ln x) - \frac{4}{x}\right\}_{x=1-3a}^{x=0.25} \tag{41}$$

Power coefficients for ideal Horizontal Axis Wind Turbine (HAWT) based on linear momentum analysis and present analysis are shown in Fig. 5.

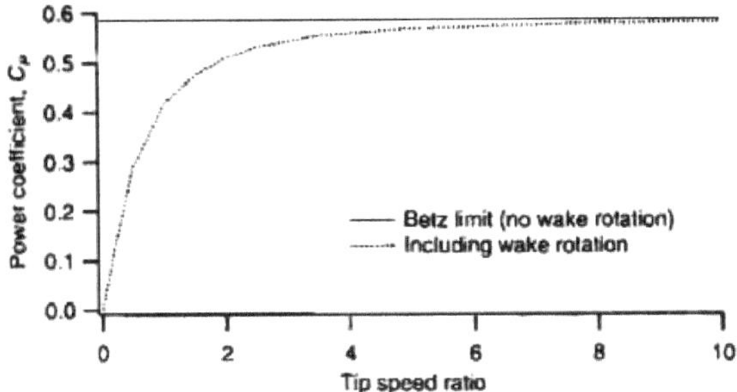

Fig. 5: Theoretical maximum power coefficient as a function of tip speed ratio for an ideal HAWT, with and without wake rotation [MMR 02]

3. Airfoils and aerodynamics concepts

The wind turbine blade has an airfoil cross-section that plays important role in the development of mechanical power. The selection of airfoil is very important for wind turbine blade design.

An airfoil is characterized by a number of terms as shown in Fig. 6. The line joining halfway points between upper and lower surfaces of airfoil is called as a mean camber line. The straight line joining the leading edge and trailing edge is called as chord line. The distance between the leading edge and trailing edge measured along the chord line is defined as the chord, c. The distance between the mean camber line and the chord line, measured perpendicular to chord line, is designated as camber. The distance between upper and lower surfaces, measured perpendicular to chord line is the thickness of the airfoil. The angle between the relative wind and the chord line is termed as the angle of attack, α [MMR 2002].

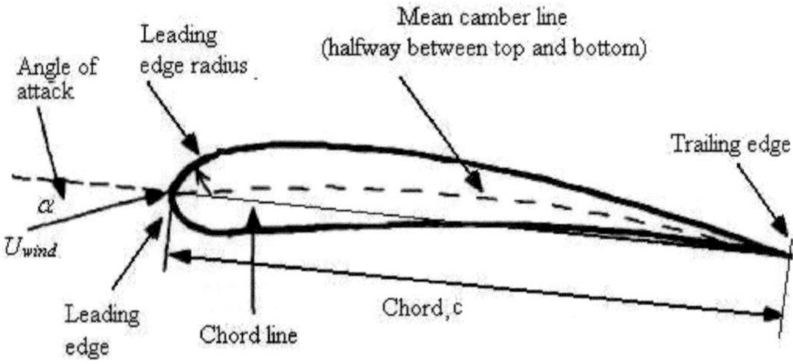

Fig. 6:   Airfoil terminology [MMR 2002]

Airflow over an airfoil generates forces over the airfoil surface. The air flow velocity increases over the convex surface of the airfoil. This result in low average pressure on the suction side of the airfoil compared to the concave side of the airfoil. In the meantime, viscous friction between the air and the airfoil surface slows the airflow to some extent next to the surface. The effect of these pressure and friction forces is presented in the form of two forces, Lift (F_L) and Drag (F_D) as shown in Fig. 7. The lift force is produced because of pressure difference on the upper and lower airfoil surfaces. This force is perpendicular to the direction of the oncoming flow. The second force parallel to the direction of oncoming flow is termed as the drag force. The drag force is the effect of viscous friction forces at airfoil surface and the pressure difference on airfoil surfaces facing toward and away from the oncoming flow [MMR 2002].

The main purpose of using airfoil section in wind turbine blade is to produce a torque about the axis of rotation. This torque is a result of circumferential force in the direct of rotation as shown in Fig. 7. From the figure it is clear that the lift force acts to increase the torque and drag force works to reduce the torque. The wind turbine performance is depending on the ratio of lift to drag, rather than the individual values [Wood 11].

The lift and drag forces are represented by non-dimensional terms lift coefficient (C_l) and drag coefficient (C_d) respectively.

$$C_l = \frac{\text{Lift force / unit length}}{\text{Dynamic force / unit length}}$$

$$C_l = \frac{l}{\frac{1}{2}\rho U^2 c} \tag{42}$$

$$C_d = \frac{\text{Drag force / unit length}}{\text{Dynamic force / unit length}}$$

$$C_d = \frac{d'}{\frac{1}{2}\rho U^2 c} \tag{43}$$

where, l and d are the lift and grad forces per unit length respectively.

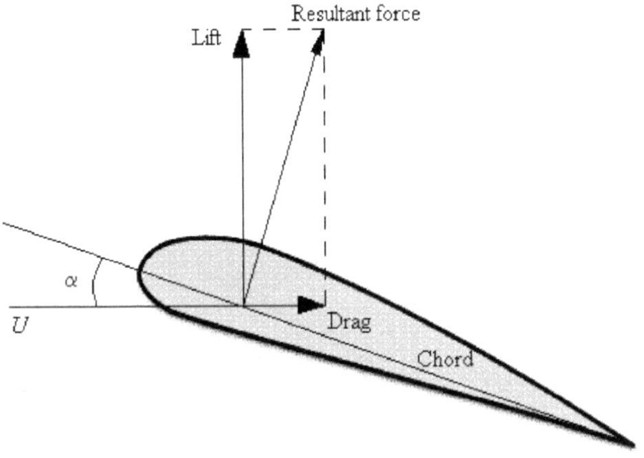

Fig. 7:   Lift and drag forces on stationary airfoil

The lift coefficient and drag coefficient are the most commonly used important parameters in blade design rather than lift and drag. The terms in these two coefficients are analogous to the most important non-dimensional parameter Reynolds number, which is used for defining the characteristics of fluid flow. The Reynolds number is defined as,

$$R_e = \frac{\text{Inertial force}}{\text{Viscous force}}$$

$$R_e = \frac{UL}{v} = \frac{\rho UL}{\mu}$$ (44)

4. Blade element theory

The forces acting on wind turbine blade can also presented in the form of lift coefficient, drag coefficient and angle of attack using blade element theory. During this analysis the blade is divided into a number of elements, N as shown in Fig. 8. In this analysis the following assumptions are made,

- No radial aerodynamic interaction between elements
- The calculations of forces on the blades are completely based on the lift and drag characteristics of the blade airfoils.

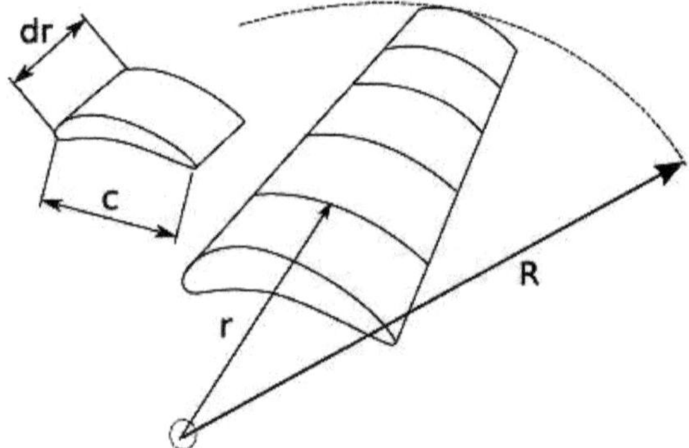

Fig. 8:   Schematic of blade element [BSJB 2001]

Consider a wind turbine with number of blades as B and tip radius R for each blade. For element at radius r, the chord length is c. The wind speed is U and blades are rotating at angular velocity, Ω. The tangential velocity, Ωr of the blade element and the induced tangential velocity of the wake, $\Omega a'r$ are shown in Fig. 9. The net tangential velocity experienced by the blade element is given by combining these two velocities.

$$\Omega r + \left(\frac{\omega}{2}\right) r = \Omega r + \Omega a' r = \Omega r \left(1 + a'\right)$$ (45)

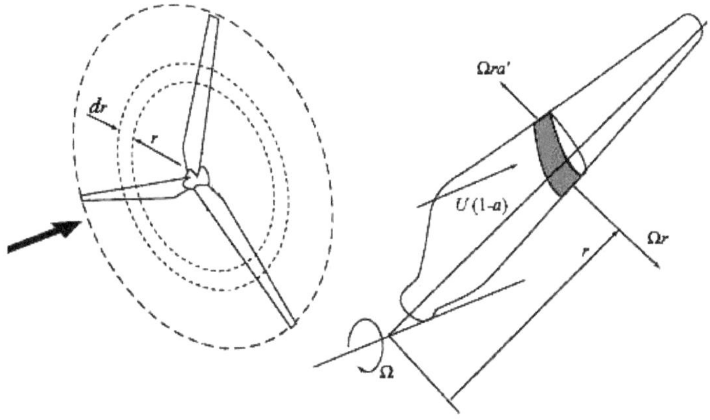

Fig. 9: Overall geometry of horizontal axis for blade element analysis [BSJB 2001]

Fig. 2.10 shows the various velocities at blade element, looking from the blade tip at radius, r. The resultant relative velocity acting at an angle ϕ is given as,

$$U_{rel} = \left[U^2 \left(1-a\right)^2 + \Omega^2 r^2 \left(1+a'\right)^2 \right]^{\frac{1}{2}} \tag{46}$$

In Fig. 10 the angle between the chord line and the plane of rotation of the rotor is known as section pitch angle or setting angle β. The angle between the chord line and the relative wind is termed as angle of attack, α. The angle between plane of blade rotation and relative wind velocity is known as angle of relative wind, ϕ. The angle of relative wind is also given as sum of angle of attack and section pitch angle.

$$\phi = \beta + \alpha \tag{47}$$

The blade twist angle, β_T is the difference between section pitch angle, β_p and the pitch angle at the tip, $\beta_{p,0}$.

$$\beta_T = \beta_P - \beta_{P,0} \tag{48}$$

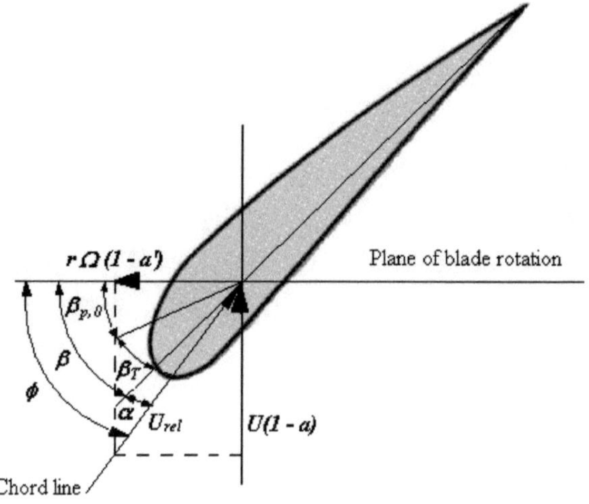

Fig. 10: Velocities at blade element

The relation between net tangential velocity and wind velocity at blades is expressed as,

$$\tan\phi = \frac{U(1-a)}{\Omega(1+a')} = \frac{1-a}{(1+a')\lambda_r} \tag{49}$$

The relative wind velocity is also given as,

$$U_{rel} = \frac{U(1-a)}{\sin\phi} \tag{50}$$

Fig. 11 shows forces acting on blade element at radius, r. In figure dF_L is the elemental lift force, dF_D is the elemental drag force, dF_N elemental force normal to the plane of rotation, and dF_T is the elemental force is the tangential to the circle swept by the rotor. The equations for elemental lift and drag forces can be written from definitions of lift coefficient and drag coefficient as,

$$dF_L = C_l \frac{1}{2}\rho U_{rel}^2 c\,dr \tag{51}$$

16

$$dF_D = C_d \frac{1}{2} \rho U_{rel}^2 c\,dr \tag{52}$$

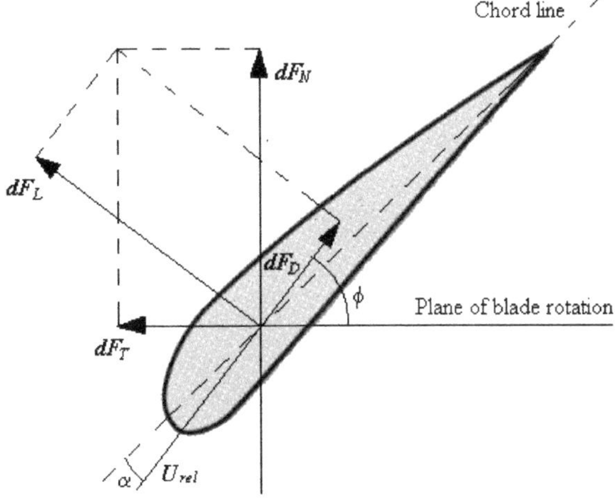

Fig. 11: Forces on blade element

From Fig. 2.11 the dF_N and dF_T are expressed as,

$$dF_N = dF_L \cos\phi + dF_D \sin\phi \tag{53}$$

$$dF_T = dF_L \sin\phi - dF_D \cos\phi \tag{54}$$

Substituting dF_L and dF_D from Equations 51 and 52, we get,

$$dF_N = B\frac{1}{2}\rho U_{rel}^2 \left(C_l \cos\phi + C_d \sin\phi\right) c\,dr \tag{55}$$

$$dF_T = B\frac{1}{2}\rho U_{rel}^2 \left(C_l \sin\phi - C_d \cos\phi\right) c\,dr \tag{56}$$

It is convenient to introduce normal load coefficient and tangential load coefficient [Han 2008] as follows,

$$C_n = C_l \cos\phi + C_d \sin\phi \tag{57}$$

$$C_t = C_l \sin\phi - C_d \cos\phi \tag{58}$$

17

After substituting C_n, C_t and U_{rel} from Equations 55, 56 and 50 respectively, dF_N and dF_T can be expressed as,

$$dF_N = \frac{1}{2}B\rho\frac{U^2(1-a)^2}{\sin^2\phi}C_n c\,dr \tag{59}$$

$$dF_T = \frac{1}{2}B\rho\frac{U^2(1-a)^2}{\sin^2\phi}C_t c\,dr \tag{60}$$

The tangential force acting on the element at radius r causes differential torque and given as,

$$dQ = B\,r\,dF_T \tag{61}$$

$$dQ = \frac{1}{2}B\rho\frac{U^2(1-a)^2}{\sin^2\phi}C_t c r\,dr \tag{62}$$

The Equations 60 and 62 will be useful to determine the ideal blade shape for optimum performance.

Defining local solidity σ as,

$$\sigma = \frac{Bc}{2\pi r} \tag{63}$$

$$dF_N = \sigma\pi\rho\frac{U^2(1-a)^2}{\sin^2\phi}C_n r\,dr \tag{64}$$

$$dQ = \sigma\pi\rho\frac{U^2(1-a)^2}{\sin^2\phi}C_t r^2\,dr \tag{65}$$

5. Blade Element Momentum (BEM) theory

Momentum theory analysis for the forces calculation considers the conservation of the linear and angular momentum in the controlled volume. In the blade element analysis forces, generated by lift and drag coefficients are calculated at an element section, as a function of blade geometry. Blade Element Momentum (BEM) theory combines the results of these two methods. This theory is also termed as strip theory in literature. This theory can be used to determine the power extracted by wind turbine rotor based on airfoil shape and generated forces.

5.1 Blade element momentum theory without wake rotation [MMR 2002]

This blade momentum element analysis is carried out to determine ideal blade shape without considering wake rotation, drag force and losses from finite number of blades at maximum value of axial induction factor (a = 1/3). In this analysis wind turbine of the radius R consisting of number of blades, B, design tip speed ratio, λ and airfoil with known lift and drag characteristics as a function of angle of attack need to be selected.

Equation 23 gives the force on annular cross- section in momentum theory analysis at radius, r and Equation 55 gives normal elemental force at radius, r as per blade element theory. Equating these two forces, in the blade element momentum theory and substituting, a = 1/3 and $C_d = 0$,

$$4a\left(1-a\right)\frac{1}{2}\rho U^{2}2\pi r dr = B\frac{1}{2}\rho\left(\frac{U\left(1-a\right)}{\sin\phi}\right)^{2}\left(C_{l}\cos\phi+C_{d}\sin\phi\right)cdr \tag{66}$$

$$\frac{C_{l}Bc}{4\pi r}=\tan\phi\sin\phi \tag{67}$$

The angle of relative wind can be determined from Equation 49 after substituting $a' = 0$ and a = 1/3,

$$\tan\phi=\frac{2}{3\lambda_{r}} \tag{68}$$

The chord of the blade element section can be calculated by combining Equations 67 and 68,

$$c=\frac{8\pi r\sin\phi}{3BC_{l}\lambda_{r}} \tag{69}$$

Also,

$$\phi=\tan^{-1}\left(\frac{2}{3\lambda_{r}}\right) \tag{70}$$

5.2   Blade element momentum theory with wake rotation

The axial induction factor, a and angular induction factor a' can be calculated from blade element momentum theory by equating equations of forces and moments derived in momentum and blade element theories [Han 2008]. Two equations are obtained as below.

$$a = \frac{1}{\dfrac{4\sin^2\phi}{\sigma'\left(C_l\cos\phi + C_d\sin\phi\right)}+1} = \frac{1}{\dfrac{4\sin^2\phi}{\sigma'C_n}+1} \tag{71}$$

$$a' = \frac{1}{\dfrac{4\sin\phi\cos\phi}{\sigma'\left(C_l\sin\phi - C_d\cos\phi\right)}-1} = \frac{1}{\dfrac{4\sin\phi\cos\phi}{\sigma'C_t}-1} \tag{72}$$

For simplification in calculation of these factors, Wilson and Lissaman [WL 1974] shows that substituting $C_d = 0$ gives results with negligible error for the airfoils with low drag coefficient. So these factors can be written as,

$$a = \frac{1}{\dfrac{4\sin^2\phi}{\sigma'C_l\cos\phi}+1} \tag{73}$$

$$a' = \frac{1}{\dfrac{4\cos\phi}{\sigma'C_l}-1} \tag{74}$$

5.3   Tip loss factor

Because of air tends to flow around the tip from lower surface to the upper surface, reducing the lift force near the tip and hence the power near the tip. A number of approaches have been developed to include the effect of tip loss. The most popular method is developed by Prandtal. The correction factor, F introduced by Prandtal is given as [Han 2008],

$$F = \frac{2}{\pi}\cos^{-1}e^{\frac{\left(\frac{B}{2}\right)\left(1-\frac{r}{R}\right)}{\frac{r}{R}\sin\phi}} \tag{75}$$

After including Prandtal correction factor, the induction factor Equations 71 and 72 can be written as,

$$a = \frac{1}{\dfrac{4F \sin^2 \phi}{\sigma' C_n} + 1} \tag{76}$$

$$a' = \frac{1}{\dfrac{4F \sin \phi \cos \phi}{\sigma' C_t} - 1} \tag{77}$$

Equations 2.73 and 2.74 can be written as ,

$$a = \frac{1}{\dfrac{4F \sin^2 \phi}{\sigma' C_l \cos \phi} + 1} \tag{78}$$

$$a' = \frac{1}{\dfrac{4F \cos \phi}{\sigma' C_l} - 1} \tag{79}$$

5.4   Blade Shape for optimum rotor with wake rotation

The angle of the relative wind and the chord are determined by Equations 80 and 81 by considering the effect of wake rotation [MMR 2002]

$$\phi = \frac{2}{3} \tan^{-1} \left(\frac{1}{\lambda_r} \right) \tag{80}$$

$$c = \frac{8\pi r}{BC_l} \left(1 - \cos \phi \right) \tag{81}$$

6.   Aerodynamic performance of tilted rotor

It is important to determine the aerodynamic performance for the purpose of energy production estimation. As rotor makes inclination with the wind flow direction it is producing less power compare to normal rotor. Hence, normal theories cannot be applied to determine aerodynamic performance. The modified theories for tilted rotor should be used. According to momentum theory applied for the tilted rotor making inclination (γ) in horizontal plane, maximum value of coefficient of performance is given by the Equation 82 [BSJB 2001],

$$C_{p,\text{max}} = \left(\frac{16}{27}\right)\cos^3 \gamma \tag{82}$$

Glauert's (1926) momentum theory modified [BSJB 2001] for tilted rotor gives maximum C_p as,

$$C_p = 4a\sqrt{1-a(2\cos\gamma-a)}\left(\cos\gamma-a\right) \tag{83}$$

By considering vortex cylinder model the power coefficient for the tilted wind turbine is given by the Equation 84 [BSJB 2001],

$$C_p = 4a\left(\cos\gamma + \tan\frac{\chi}{2}\sin\gamma - a\sec^2\frac{\chi}{2}\right)\left(\cos\gamma-a\right) \tag{84}$$

A comparison for the maximum C_p values derived from above theories, as a function of shaft inclination, is shown in following Fig. 2.12

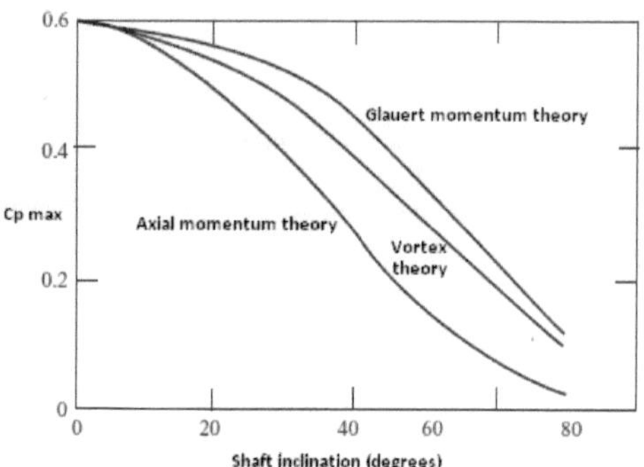

Fig. 12:   Change in coefficient of performance with shaft inclination [BSJB 2001]

REFERENCES

[BSJB 2001]   **Burton Tony, Sharpe David, Jenkins Nick, Bossanyi Ervin**, Wind Energy: Handbook, John Wiley & Sons, Ltd, England, 2001

[Han 2008]   **Hansen Martin O. L.**, Aerodynamics of Wind Turbines, Second Edition, Earthscan in the UK and USA in 2008

[Joh 2006]   **Johnson G. L.**, Wind Energy Systems, Kansas State University, First edition, 2006

[Kha 2006]   **Khan B. H.**, Non-Conventional Energy Resources, Tata McGraw-Hill Publishing Company Limited, New Delhi, 2006, pp.173.

[MMR 2002]   **Manwell, J. F. McGowan J, G. and Rogers A, L.**, Wind Energy Explained: Theory, Design and Application, John Wiley & Sons Ltd, England, 2002

[Pat 2006]   **Patel Mukund R.**, Wind and Solar Power Systems: Design, Analysis, and Operation, Second Edition, CRC Press, Taylor & Francis Group, NW, 2006

[Wood 2011]   **Wood David**, Small Wind Turbines: Analysis, Design, and Application, Springer-Verlag London Limited, 2011

YOUR KNOWLEDGE HAS VALUE

- We will publish your bachelor's and
  master's thesis, essays and papers

- Your own eBook and book -
  sold worldwide in all relevant shops

- Earn money with each sale

Upload your text at www.GRIN.com
and publish for free